# The Missing Scientific Essays

## Miguel A. Sanchez-Rey

# Table of Contents

The Armageddon Scenario:  War-Game


[Author:  Miguel A. Sanchez-Rey]

A Class 2 civilization has transitioned into a peaceful stage in the form of the self-management of democratic political economy. Capable of traveling in and out of wormholes and utilizing star-gates human civilization remains far apart from the Scientific Age. Having reached the Advance Age humanity remains an independent civilization with a strong-sense of independence and in which, having lost contact with other space-habitats and terraform planets, it still very much endeavors as a cooperative society of economic trade and competition between a federation of industrial bodies. This, in all, coincides with the capacity to harness the energy of a star. By harnessing the energy of a star a Class 2 civilization is able to achieve feats unheard off to human history. In which human beings has achieved a Pangea on Earth and other habitats. Other civilizations remain independent of human history and human history is very much independent of other civilizations in the rising formulation of an internationalist model. As the centuries go by human civilization travels far into the distance to harness more star-systems careful not to encroach on other independent planetary systems that lay claim to their own solar-habitats.

Within the habitable zone of the Milky Way man-kind must expand further to keep up with rising population growth and to provide for the self-management of labor that requires the ever increasing need for energy and mineral resources. Other civilizations, in which co-habitation isn't possible, also strive to reach outward their own territorial claims.  At a certain point the need becomes to harness many stars, and if then other civilizations, being independent on their own terms but forming an alliance, become in dire need of resources.  Since greater population requires, through the laws of biological and synthetic thermodynamics, more resources to survive and adapt, a point of no return becomes more ominous: that is a drought begins to emerge that begins to dwindle the availability of stars at an accelerated rate, in which the lack of even oxygen becomes more tantamount causing the eruption of a galactic conflict that competes for ever increasing dwindling resources.

That said the point of no return has been reach, and in that manner human civilization collapses and other star systems, even primitive star systems, become embroiled in a large-scale galactic conflict.  In which the galaxy has reach the dawn of the Incalculable Age:  an age with the potential to achieve the impossible but also an age in which there is the potentiality for enormous grief where all things in the end are bound to be forgotten.

The resolution to the crisis of the dawn of the Incalculable Age is the formulation of Incalculability.  A mathematical finite algebraic structure that far surpasses SUPREME and in which its development may only be possible within the next two centuries.  At its earliest stages it means the achievement of the impossible and in which an infinite amount of energy is not the resolution.

The Omega-Kardeshev Scale


Miguel A. Sanchez-Rey

The Kardeshev scale measures a civilizations technological progress by utilizing the Planck energy-range. There are four energy-scales of particular importance: they are referred to as Class [Type] 0/I/II/III energy-scales. Or what one calls a Class [Type] 0/I/II/III civilization. Where one stands at the moment is at the scale of .7 in which by 2100 one reaches Type 1. A Type 0 civilization is at the infancy of technological development. As one gets closer to Type 1 one achieves a planetary society consistent with technological advances that include the world-wide web, ocean cities, advance robotics, controlling the weather, and nuclear fusion. That is by the time one reaches Type 1 one has harness the energy of the hydrogen atom, solar-energy, geo-thermal energy, and etc. Class [Type II] civilization has harness the energy of a star, with technological advances in star-gates and time-travel, and Class [Type III] civilization has harness the energy of many stars on a galactic scale in which nothing known to science can destroy a Class III civilization.

PHPR [The Physicalist Program] is design as a resolution to a foreseeable catastrophic scenario in the Scientific Age in the form of task. The First Task is a 100 Year Task. The First Task aims to achieve a terraformic reaction that will resolve the long-standing issue of mineral depletion. Coinciding with ITER [International Thermonuclear Experimental Reactor] an integrationist endeavor, with gaining access to metaspace, will culminate in completing The Grand

Unification Scheme at the Omega-Energy Scale. The Omega-Energy Scale eludes to a slight-shift in human-technological development that coincides with the dawn of the Scientific Age. That is the new energy-scale that revises the Kardeshev scale is the Omega-Kardeshev scale.

The Omega-Kardeshev scale measures a civilizations technological progress by utilizing the same methodology as the Kardeshev-scale only that with The Grand Unification Scheme will completion yield technological advances that are unforeseeable. Unforeseeable in that one has a closer understanding of what a civilization may potentially be like and not just what a civilization looks like. All subsequent tasks, after the completion of The First Task, will achieve rapid technological advances which will conclude with the last task at the dawn of the Advance Age -- inferred as a Class [Type] II civilization. At the Incalculable Age one has reach Class [Type] III civilization where certain technological advances will be achieved much earlier than expected; that is, Class II. The ages become longer because of space-travel and the vast distances of the cosmos.

The interplay between the Omega scale and the Kardeshev scale gives one an idea between what a civilization may be like and what a civilization may potentially look like. In that manner applying further progress in PHPR will achieve perfection of the Omega-Kardeshev scale that will gain greater value for anticipatory studies in sociology and psychology.

# Incalculability

## Miguel A. Sanchez-Rey

The Physicalist Program

## Abstract

Define and explain Incalculability.

April 17th, 2017

Isaac Newton define calculus to be the study of measures and numbers. That is calculus is understood to involve the differential and integral form in an attempt to evaluate the physical phenomenon of motion and energy. Since then calculus has spawn whole new mathematical fields in such areas as geometry, topology and analysis. Furthering advances in such physical phenomenon as electrodynamics and thermodynamics. Though trivial in that sense Isaac Newton brought numbers to the field of the physical sciences so that quantifiable results can be made that validate the interplay between theory and experiment. Once such validation is made then one achieves natural law.

The evolution of theoretical mathematics has shown that advances in the physical sciences yield further advances in the mathematical sciences. Even then an interplay surmounts which is known to be the field of mathematical physics.

In the field of mathematical engineering calculus, or what is understood to be algebraic and differential topology, allows one to quantify results to achieve a giving task that leads to advances in the technological sciences.

PHPR [The Physicalist Program] has further one's understanding of SUPREME. A mathematical operator that is a homogenous topology. Giving this homogenous topology one can control and manipulate both entropy and the conservation of energy which then integrates with computational control. How then should SUPREME achieve so much more if it's to be design as a property of control measure theory.

It's an abstruse object. An object that has been exhausted but still very much eloquent in design and detail.

---

Define: Incalculability:
$$[ \quad ] \ \llcorner \ [ \quad ] \rightarrow \ | \quad |$$

---

As the functor between both a base and target homogenous topological space to a morphic topological space. Whereby a morphic topological space is intrinsic to any sets besides the null set and empty set. That is both a target and base homogenous topological space can morphologically map onto a topological space that is besides the null set and empty set. In other words, given any number space in a homogenous topology the morphic topological space can take any value that is intrinsic to the parameters of the number space. The parameters of the number space being the ∟ - parameter.

That said the initial standpoint behind Incalculability is the capacity to achieve properties of any quality giving the qualities that are presented that exist in motion and energy. The ∟ - parameter helps to achieve control of any construction; that, is measure theory is embedded as a limit of what is available in terms of any number or any other kind of quantifiable sets.

The Mosaic of the Norm


[Author:  Miguel A. Sanchez-Rey]

The sociopathic norm sets the norm in a very frightening way: by making flaw decision-making that causes havoc in such a way that such flaw decision-making becomes the norm. Academic and humanitarian frauds that resort to both academic and financial fraud to achieve and sustain their fame and fortune. Seeking advantage over others by pursuing fraudulent means to accelerate their writing methods and to increase their wealth.

They are nevertheless conceptualized as a People's Temple Cult: a brutal crime-ring that spans world-wide but can also be seen as a group of individuals that share interrelated lives though remain unaware of their interrelationships to each other and how much they depend on each other to preserve their power and influence. Some members of the norm sustain a small-following and others have a large-following. But what they have in common is that they are horrid.

By horrid one means capable of horrific acts of hatred and malice but in their public lives they present themselves as loving creatures with messianic ambitions. They live fraudulent life-styles that they nevertheless transcend themselves as actors in a large mosaic that carry on with their personal lives pursuing their responsibilities and goals oblivious to the other lead actors that interrelate with their own lives. They can be even said to be have an engrained view of themselves and those they interrelate with. That when confronted with each other it nevertheless ends in murder and genocide. They are, nevertheless, a horrific form of a Cloud-Atlas mosaic.

A norm that has cause world-wide havoc, and that has seen the dawn with the Windows XP operating system, is a short-live cult phenomenon of flaw decision-making but nevertheless a world-wide threat in which groups of actors pursue their own ends without consideration of what those entail besides the desire for fortune and fame. Cult-figures posing as academic and financial experts they amass a large-following whom nevertheless remain complacent in the standards of the mosaic of the norm.

A 100 Year Task That Involve Seven Impossible Task

Miguel A. Sanchez-Rey

Seven Impossible Tasks were set in the groundbreaking textbook, *The Physicalist Program* that laid the bases for a century-long task to complete The Grand Unification Scheme that will concurrently ignite and take advantage of the terraformic process. Since then the physical sciences have undergone significant progress in high-energy physics and mathematical physics. There has been advances in different specialized fields such as quantum computers, solid-state physics, supersymmetry, super-quantum cosmology, and etc. Here one delves on the primary advances that has significance to PHPR [The Physicalist Program].

It's discerned that the planet is undergoing an energy crisis that coincides with a climate crisis. To forestall this crisis requires a new energy-source. One that is clean and efficient. That can produce enough energy to power the planet and one in which the byproducts of such energy-source won't do any further damage to the Earth's climate. As well such energy-source must keep up with rising population growth and advances in high-tech. The only clean-energy source capable of meeting those demands is understood as the International Thermonuclear Experimental Reactor [ITER], in France, that aims to impart energy that is equivalent to the amount of energy produce by the sun.

Clean and efficient that no lasting damage will be done to the environment, or the Earth's climate, and in which environmental recovery begins to take form at a global scale.

That is by fusing hydrogen atoms utilizing a unique design of state-of-art lasers, and by resolving the problem of confinement, ITER finally achieves, at a large-scale, a nuclear fusion reaction.

But clean-energy comes at a heavy price: mineral depletion. More energy output from ITER means higher economic growth. Hence higher industrialization and manufacturing but less inflation. That is at a certain point minerals become scarce. With rising mineral scarcity comes rising food-scarcity due to an agricultural deterioration in the use of soil and other resources needed to sustain agriculture. As well higher-prices on goods and services due to dwindling manufacturing productivity cause by a scarcity in needed minerals to produce electronics, shoes or other essential items to sustain a global economy. Where regional territorial conflict and displacement becomes more ominous and deadly.

In that manner one undergoes a domino effect that eventually leads to a planetary decline consistent with a rotting planet -- a dead planet.

PHPR establish The First Task as a resolution to a foreseeable catastrophic scenario in the Scientific Age in the form of a task. The First Task has been set as the resolution to mineral depletion. That is to ignite a terraformic reaction, and to take advantage of the terraformic process, by simultaneously completing and utilizing The Grand Unification Scheme.

Seven Impossible Tasks has been install that aim to organize PHPR's endeavor in completing The First Task.  60 percent of The First Task must be completed within the 40-year time frame and 40 percent must be completed within the next 60-year time frame which means a 100 Year Task has been set.  After the 40-year window of opportunity ITER will begin to be mass-produce in which global industrial development begins to rapidly accelerate and in which a global economic, and environmental decline, starts to slowly manifest.  Allowing a 60-year time frame to carefully reach completion of The First Task.  Gaining access to metaspace approximately 20-years after ITER goes online.

There has been ongoing controversy about the 1st impossible task.  That is dark matter and supersymmetry.  It's known that current particle physics experiments has produce little if not no tangible evidence of dark matter.  Implying that dark matter doesn't exist or that current knowledge of dark matter may have to be modified.

It's expected that the hierarchy problem will be abandoned in favor of metaspace.  Then experimentally proving the existence of dark matter faces a significant obstacle -- as hierarchy is essential for dark matters existence.  If SUSY-like physics resides in metaspace then one's notion of dark matter will change.  But is dark matter essential for completing The First Task?

Dark matter exists as WIMPS that weakly interact with ordinary matter and since inducing a terraformic reaction aims to control matter at the quantum scale then one can drop dark matter even if whether or not significant modifications are to be made. That said supersymmetry still resides as SUSY-like physics in metaspace.

Accounting for all 121 variant [of stringy]'s, stipulated by the TrH Theorem, as originally proposed, is now an attempt to gain access to metaspace. Whereby computational control and SUPREME has undergone advances of their own. SUPREME is of utmost importance and in which progress in SUPREME has been exhausted. That is to indefinitely sustain the terraformic process means enforcing Incalcubility, which naturally integrates with computational control, and as a dynamical property of SUPREME, it will also be use, along with computational control, to harness metaspace that eventually, with prime factorization, leads to controlling and manipulating the terraformic process through completion of The Grand Unification Scheme.

Casting doubt on the AdS/CFT correspondence whereby a limitless measure has been applied to control holography.

The completion of The Grand Unification Scheme becomes of imperative importance and in which miniscule revisions can be made as one arrives closer to completing The First Task:

A 100 Year Task That Involve Seven Impossible Task

1$^{st}$ Task:  SUSY and ITER.

2$^{nd}$ Task: Simplification of the Physics.

3$^{rd}$ Task: Metamorphic space.

4$^{st}$ Task:  The mathematics will change since computation and probability has been eliminated.

5$^{st}$ Task:  Computational Control and Incalculability.

6$^{st}$ Task:  Holography toward the Grand Unification Scheme; you can make revisions.

7$^{st}$ Task:  You can only make miniscule revisions.

Where simplification of the physics is to continue by imposing Logical Form [LF].

In which one dares not get carried away by prime factorization.

# PHPR Preconditions

The Physicalist Program [PHPR] top-scientists are arranged with preconditions design to protect them during a set task. When a task is near-completion preconditions are implemented.

When a task is safely completed…clear implies denied.

When a task leads to failure or loss of communication…denied implies confirm.

------------------------

Obey the International Criminal Court of Justice at the Hague.

The United States of America Armed Forces and the North Atlantic Treaty Organization Armed Forces has full ownership of PHPR.

War-Crime… … [denied, confirm]

PHPR is top-secret.

Procedural Protocol has been set…

The United States Ivy Leagues permanently shut-out…

The Microsoft Corporation permanently shut-out…

The Business Private Sector permanently shut-out…

The Royal Society of London for the Improvement of Natural Knowledge is permanently denied security clearance…

The Nobel Prize Committee in Physics is permanently denied security clearance…

The Nobel Prize Committee in Chemistry is permanently denied security clearance…

The Nobel Prize Committee in Peace is permanently denied security clearance…

The Abel Prize Committee is permanently denied security clearance…

St. Mary's, University at Oxford on standby for assistance…

Trinity College, University of Cambridge on standby for security clearance…

… [clear, denied]

We have no knowledge of the Scientific Age.

www.ingramcontent.com/pod-product-compliance
Lightning Source LLC
Chambersburg PA
CBHW081317180526
45170CB00007B/2745